AF270690

CHEETAHS

by Elizabeth Andrews

Cody Koala
An Imprint of Pop!
popbooksonline.com

This book is filled with videos, puzzles, games, and more! Scan the QR codes* while you read, or visit the website below to make this book pop.

popbooksonline.com/cheetah

*Scanning QR codes requires a web-enabled smart device with a QR code reader app and a camera.

abdobooks.com

Published by Pop!, a division of ABDO, PO Box 398166, Minneapolis, Minnesota 55439. Copyright ©2025 by Abdo Consulting Group, Inc. International copyrights reserved in all countries. No part of this book may be reproduced in any form without written permission from the publisher. Cody Koala™ is a trademark and logo of Pop!.

Printed in the United States of America, North Mankato, Minnesota.
102024
012025

THIS BOOK CONTAINS RECYCLED MATERIALS

Cover Photo: Getty Images
Interior Photos: Getty Images, Shutterstock Images
Editor: Grace Hansen
Series Designer: Neil Klinepier, Candice Keimig

Library of Congress Control Number: 2024938582

Publisher's Cataloging-in-Publication Data
Names: Andrews, Elizabeth, author.
Title: Cheetahs / by Elizabeth Andrews
Description: Minneapolis, Minnesota : Pop!, 2025 | Series: Big cats | Includes online resources and index
Identifiers: ISBN 9781098246884 (lib. bdg.) | ISBN 9781098247447 (ebook)
Subjects: LCSH: Big cats--Juvenile literature. | Wildcat--Juvenile literature. | Cheetah--Juvenile literature. | Hunting leopard--Juvenile literature. | Cheetah--Behavior--Juvenile literature.
Classification: DDC 599.759--dc23

Table of Contents

Chapter 1

What Is a Cheetah?

A cheetah is a large cat with a long, thin body. Cheetahs have pale, yellow-brown fur and black spots. Their long, dotted tails are darker in color at the tips.

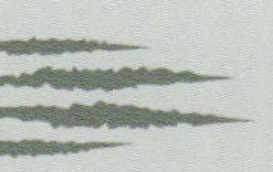
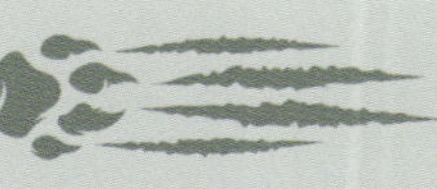

Watch a video here!

Cheetahs have excellent eyesight. They have special black markings around their eyes. The markings look like tear stains. Scientists believe these markings protect a cheetah's eyes from the Sun.

Cheetahs are the fastest
land animal on Earth! They
can run 70 mph (113 kph)!

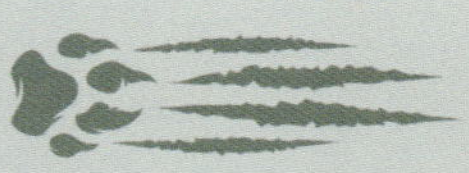

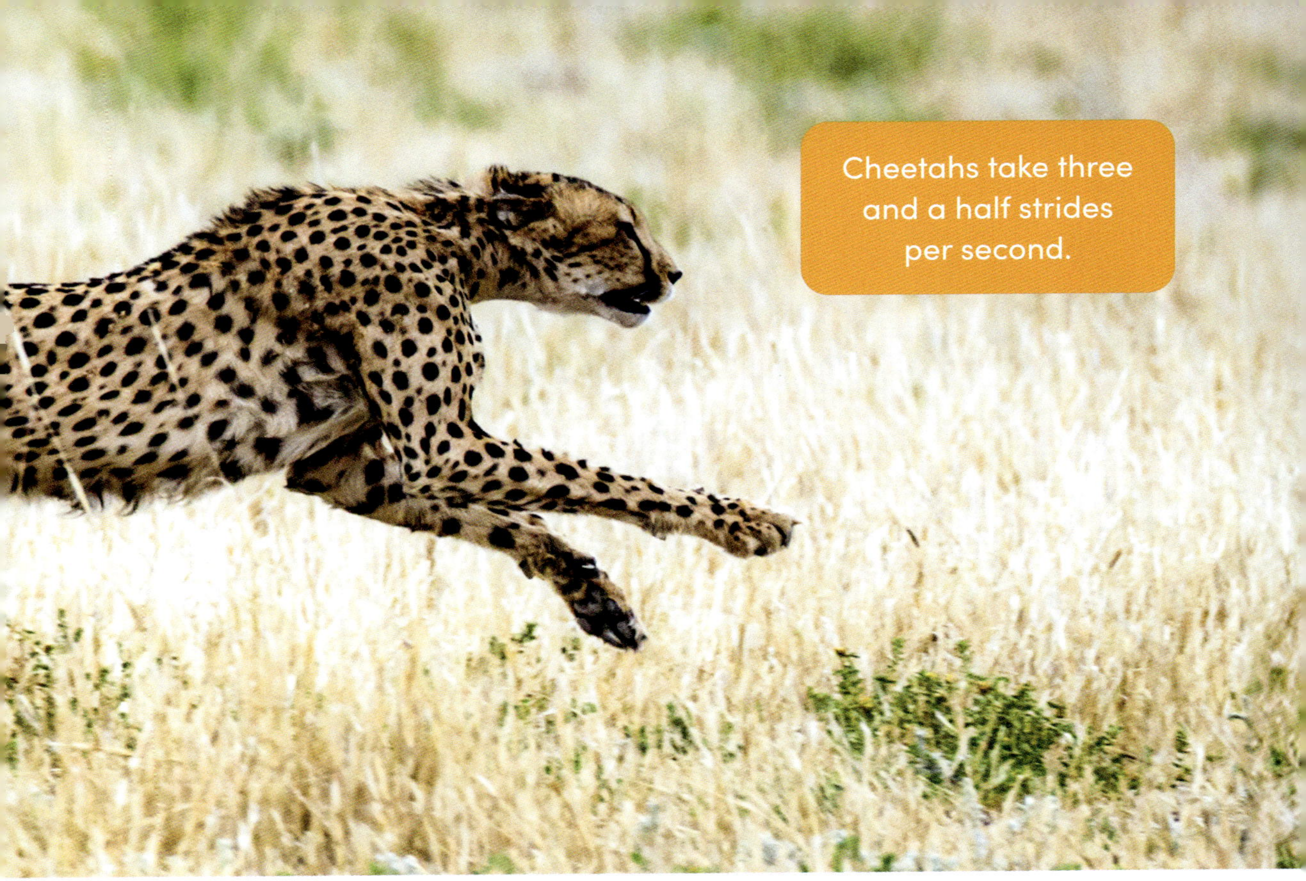

Cheetahs take three and a half strides per second.

Cheetahs long legs and spines help them run fast. Their long tails help them **steer**.

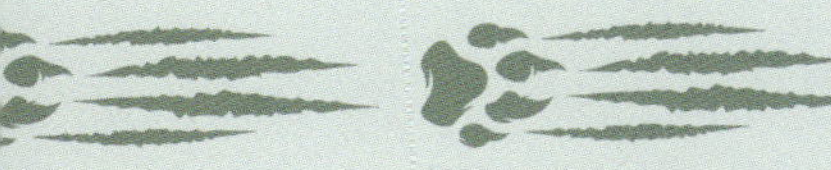

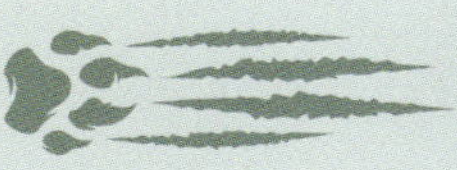

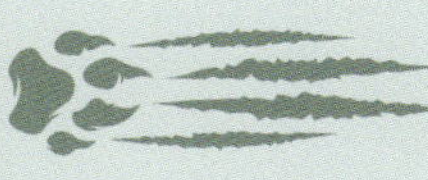

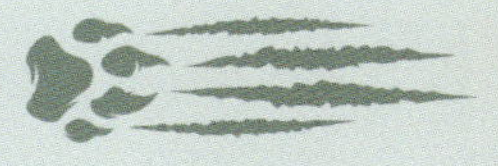

Where Do Cheetahs Live?

Cheetahs once lived throughout Africa and Asia. Over the last 50 years, they have experienced extreme **habitat** loss. Cheetahs are now only found in the **savannas** of Africa.

Where Cheetahs Live

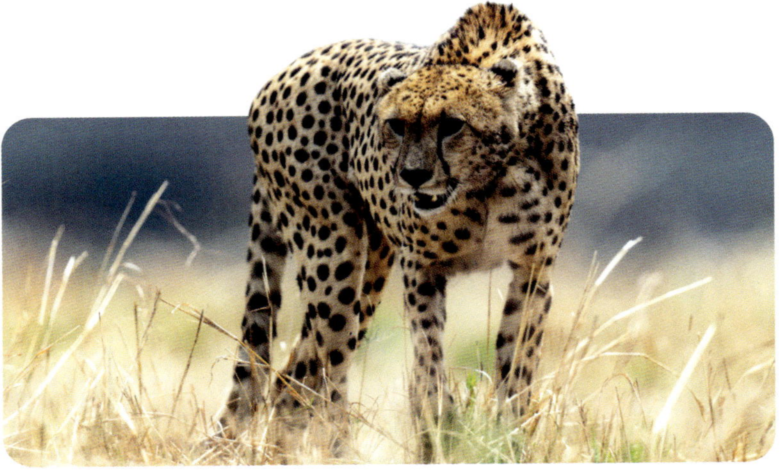

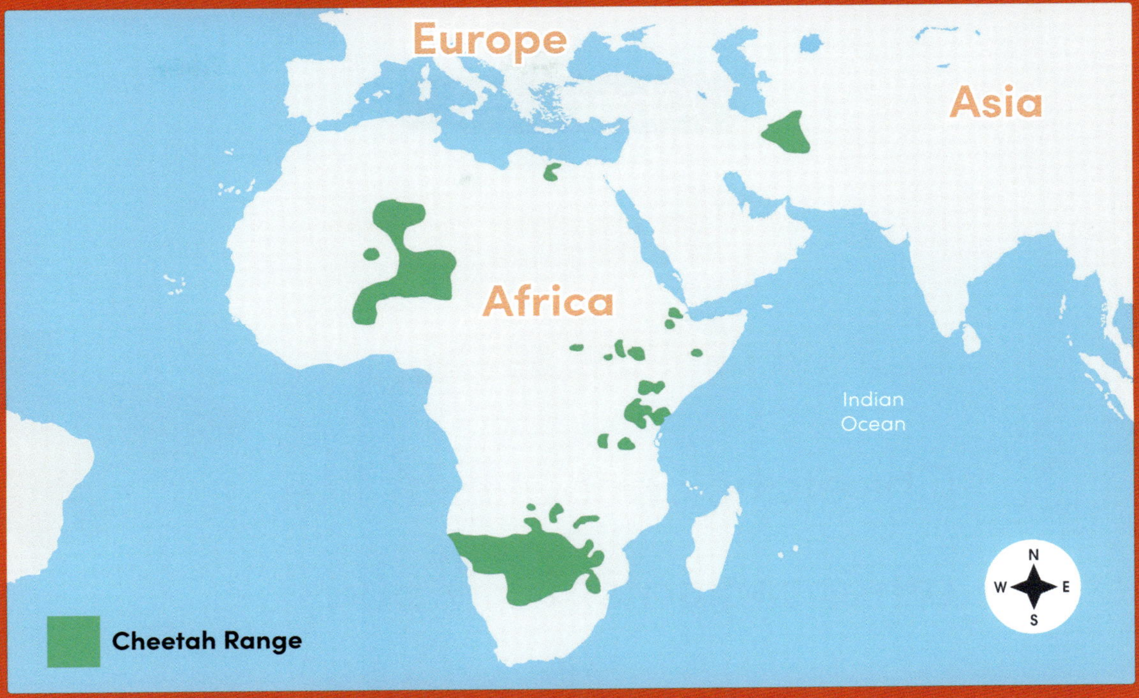
Europe
Asia
Africa
Indian Ocean
N
W
E
S
Cheetah Range

Learn more here!

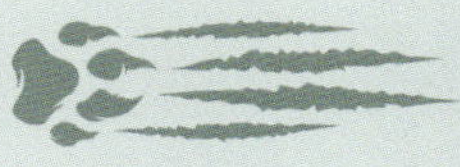

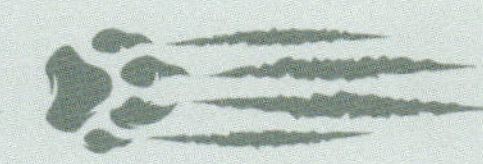

Chapter 3

Cheetah Habits

Cheetahs cannot roar like other big cats. They purr, hiss, chirp, and growl to communicate. Cheetahs are **carnivores**. They eat small- to medium-sized animals such as hares and gazelles.

House cats and cheetahs make many of the same noises.

Cheetahs are amazing
hunters. They use their speed
to hunt during the day. They
stalk an animal and then run
the last few hundred yards

The number animals cheetahs hunt is shrinking due to habitat loss. This makes it harder for cheetahs to survive.

and pounce on it. Cheetahs often must rest for about 30 minutes to catch their breath before they eat.

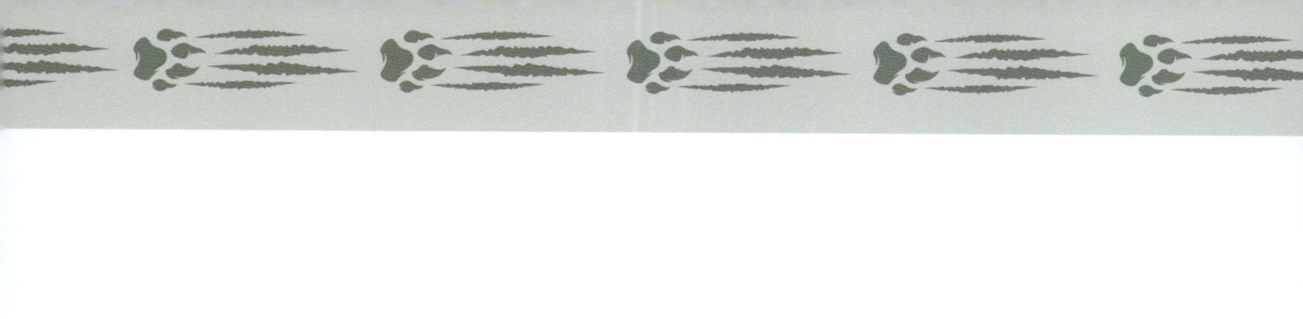

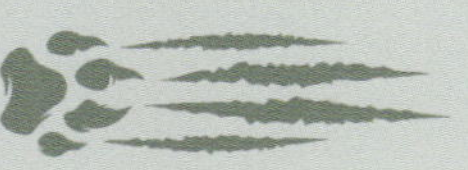

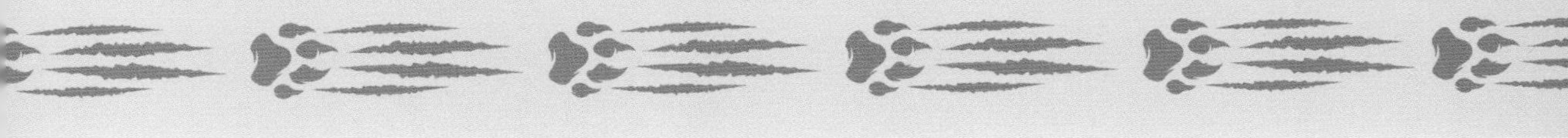

Cheetahs live in small families. Females live with their cubs. Males often live with two to three other males. They are usually brothers. They will live and hunt together for their entire lives.

Chapter 4

Cheetah Cubs

Mother cheetahs give birth to two to five cubs at a time. Cheetah cubs are born with a long, fluffy hair on their backs. This helps them hide in long grass.

Complete an
activity here!

Cubs are in danger from **predators**. Mother cheetahs move their cubs to new hiding spots every day. Cubs stay with their mothers for nearly two years. Cheetahs usually live ten to 12 years.

Male cubs leave to find their own territory. Female cubs stay in their mother's territory.

Making Connections

Text-to-Self

What big cat are you most interested in? Please explain your answer.

Text-to-Text

Have you read about any other kinds of big cats? If so, how were those big cats similar to or different from cheetahs?

Text-to-World

The cheetah is the fastest land animal. What are some other fast animals from around the world?

Glossary

carnivore – an animal that feeds on other animals.

habitat – the home of an animal.

predator – an animal that lives by hunting and eating other animals.

savanna – flat grasslands with scattered trees and shrubs.

stalk – to follow an animal as closely as possible without being seen or heard.

steer – to cause to move in a certain direction.

Index

Online Resources

popbooksonline.com

This book is filled with videos, puzzles, games, and more! Scan the QR codes* while you read, or visit the website below to make this book pop.

popbooksonline.com/cheetah

*Scanning QR codes requires a web-enabled smart device with a QR code reader app and a camera.

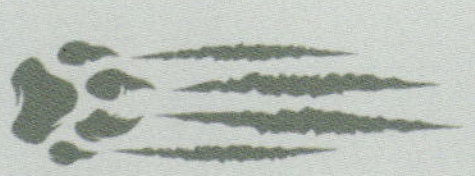
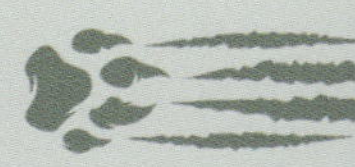